Up in the Air: European Union and Transatlantic Defence Industrial Cooperation

John Appleby and Edward Foster

Royal United Services Institute for Defence Studies

First Published 1993

© Royal United Services Institute for Defence Studies

ISBN 0-85516-089-6
ISSN 0268-1307

The Royal United Services Institute for Defence Studies (RUSI) is a professional body based in London dedicated to the study, analysis and debate of issues affecting defence and international security.

Founded in 1831 by the Duke of Wellington, the RUSI is one of the most senior institutes of its kind in the world which, throughout its history, has been at the forefront of contemporary political-military thinking through debates, public and private seminars, conferences, lectures and a wide range of publications. The independence of the Institute is guaranteed by a large, worldwide membership of those people and organisations who have a serious and professional interest in the thorough and objective analysis of defence and international security.

Critical and acclaimed analysis of issues of moment has underwritten the RUSI's Whitehall Papers for many years. The new series will, in its revised A5 monograph format, continue to provide expertise in the field. The series, which will comprise six publications a year, will address the major areas of current interest.

Whitehall Papers are available as part of a membership package, or singly at £6.50 plus p & p (£1.00 in the UK/ £2.00 overseas). Orders should be sent to the Publications Department, RUSI, Whitehall, London SW1A 2ET and cheques and postal orders made payable to the RUSI.

Printed in Great Britain by Sherrens Printers, Units 1 & 2, South Park, Granby Industrial Estate, Weymouth, Dorset.
The Royal United Services Institute for Defence Studies, Whitehall, London SW1A 2ET. Registered Charity No. 210639

CONTENTS

Preface   1

Introduction   3

From the Single Market to European Union   5

The Grand Design   17

Trade Blocs and GATT   29

Defence Industries and the Single Market   35

1993 and Onwards   47

i

NOTE ON THE AUTHORS

John Appleby and Edward Foster are Researchers in the Studies Department of the RUSI.

PREFACE

Barely three years ago, fortune, fame and luck appeared to favour the old European continent. A decades-long ideological division had been overcome not through war, as everyone had feared since 1945, but actually with the communist system being torn down by those who suffered under its rule. The Berlin Wall collapsed, and in the West other barriers were coming down peacefully. The Single European Market, a project begun in 1985, was proceeding well, and despite initial hesitation, the European Community (EC) was awakening from a hibernation which lasted for most of the late 1970s and early 1980s. The end of communism appeared, at least initially, to offer even bigger opportunities. All of a sudden the largest market the world had ever seen was a certainty. EC leaders were preparing for their Maastricht Treaty on European Union, a project arresting in its importance. All border controls would be swept aside; all barriers to trade would go. Fuelled by cheap labour from the East, supported by lower military budgets, spurred by rising expectations and boosted by a good capital structure, developed infrastructure and a large, talented and educated population, the sky was the limit for Europe.

Thus in 1990 the RUSI brought out Whitehall Paper 6, entitled *1992: Protectionism or Collaboration in Defence Procurement*. Much of its substance remains valid three years on, since the process of streamlining it describes in the defence industries predates the Single European Act and continues without strict regard to the new competition affecting the wider market.

However, since publication of Whitehall Paper 6, the language of Europe's institutions has undergone change, with the Maastricht Treaty becoming the conventional measure of the success or failure of the European Union. The EC, or 'Brussels', has thus been tempted to lift its gaze beyond the operation of a functional unified market. The WEU has seemed to advance and recede, but remains technically poised to assume a higher role in accordance with the Maastricht provisions. The Independent European Programme Group (IEPG) has been absorbed into its new structures. Most conspicuously of all, NATO—the West's source of strength in the Cold War—is undergoing an uncertain transformation, including exploring new relationships through the expedient of the North Atlantic Cooperation Council (NACC). But the most profound change, one which justifies a new analysis, is that affecting the landscape in which these organisations are sited. The Cold War's end did not simply, as widely expected, lead to a unipolar, single-superpower world. The disappearance of the Soviet military threat has encouraged Americans and Europeans alike to lower their vision from the overriding priority of collective security to focus instead on issues which divide rather than unite them. Moreover, in the absence of this outside menace, the lingering nationalisms of Western Europe have been exposed, exacerbated by the resurfacing of uglier antagonisms in Central Europe and the Balkans. The West's structures of mature democracies have, to stretch a point, been infected by the same rot that destroyed their Eastern counterparts.

This paper should therefore be seen as a companion to its predecessor, updating where necessary and building upon arguments already advanced. The speed at which events are moving (everywhere, it seems, other than in the GATT negotiations) has caused the text to be reworked more than once during preparation; compared with the relative stasis of the Cold War period, this volatility will be an unavoidable handicap for all researchers in international affairs for the foreseeable future. Three years ago, it was possible to paint a picture of a future Alliance of twin political and trade blocs; three years from now, commentators are unlikely to be pronouncing this consolidation near its end.

INTRODUCTION

This analysis starts from the unstartling assumption that transatlantic relations are polarising. The development of the European Community provides the most obvious evidence of this, more recently partnered by the emergence of a North American Free Trade Area (NAFTA). The economic pressures of open markets, to say nothing of the external competition from developing societies, is driving a process of rationalisation in each of these. NATO is being recast as an alliance resting on twin and complementary pillars, in accordance with the parallel belief that, as the Europeans consolidate their latent collective political force, they should assume greater responsibility for the bearing the burdens of their own security. As this dynamic has gathered pace, the temptation has grown—in both Europe and North America—to speak of 'European' views and interests. Community legislation and proposals obviously do this with an eye to the future, and in the context of transatlantic trade talks, the European Commission clearly advances a position agreed by prior negotiation among the EC's 12 member states. Yet the practice of extending the use of this shorthand so as to refer to the aims of 'European' industry, still more the record of 'Europe' in the Gulf and the former Yugoslavia, is frequently misleading and sometimes dangerous. Europe has never yet been a unitary entity and, despite the fears raised by the Maastricht Treaty of a 'federal superstate', shows no sign of becoming one in the near future.

Indeed this is Europe's essential problem. In NATO and the G7, there exists no European bloc or universal European position. In fact events within these bodies and the European Community are dictated by leading states pursuing often mutually conflicting national interests from appropriate platforms. Francis Fukuyama's declaration of the end of history has been judged premature in the light of the tenacious survival of conflict, and it seems quite appropriate to add in the same breath that the European nation-state remains nearly as far from extinction. What does remain a valid issue for debate, however, is the speed and degree of

3

Europe's convergence, and the strategic and commercial implications of an uncertain evolution.

FROM THE SINGLE MARKET TO
EUROPEAN UNION

It is worth recalling that, for all the unease displayed by Americans at the prospect of Fortress Europe ejecting their trade and influence, the process of integration accords with long-held US policy aims for the continent. Since the drive to the Single Market represented a response to the European Community's lack of direction during the 1970s, it was encouraged by Washington as further cementing the integration process and as offering a boost to Europe's ability to carry a greater share of the weight of the Alliance's defence. In the past, America had been prepared to build up the shattered economies of Western Europe and the developing world, overlooking technical issues of strict commercial reciprocity for a wider security objective. In the same spirit, US officials could declare 'EC 92 is user-friendly'.[1]

Moreover, the Single European Act was not simply concerned with the mechanics of felling obstacles to free trade and movement, harmonising taxation and empowering directives on vegetable specifications. In the vein of declarations made within the looser intergovernmental process of European Political Cooperation (EPC), the SEA stated that 'closer cooperation on questions of European security would contribute in an essential way to the development of a European identity in external policy matters'.[2] Similar undertakings to work for 'concrete progress towards European unity',[3] emphasised by greater endeavours towards effective foreign policy coordination[4] would later return to haunt Margaret Thatcher. In the joint Declaration on US-EC Relations of November 1990, it was further declared that 'the European Community is acquiring its own identity in economic and monetary matters, in foreign policy and in the domain of security', even though the last of these activities remain in practice intergovernmental and thus the purview of the wider European Union rather than the Community as such.

By 1990, of course, issues of foreign and security policy had been overtly pushed up the agenda of European business, by events not even dimly foreseen at the time of the framing of the Single European Act. For, with the crumbling of the Cold War order, came the reappearance of a Germany whole and free, a development which so upset Western Europe's internal balance that the more distant agenda for European Union was brought forward. For President Mitterrand and Chancellor Kohl, rapid European union amounted to a desperate, accelerated culmination to their national policies before the landscape changed; to the European Commission and its President Jacques Delors, (to whom credit goes for reinvigorating the Community in the 1980s) the leap towards Union would be the consolidation of all previous achievements. Other countries had smaller national axes to grind, creating the intricate and unwieldy chain of interlocking conditions to emerge from the Maastricht Inter-Governmental Conference of December 1991. Only in the case of the United Kingdom and Denmark were the demands essentially negative, rearguard actions to protect the remaining areas of national sovereignty and of the *status quo* in defence.

The result went far beyond the economic rationalisation yet to be effected in the Single Market, decreeing instead a corporate Europe with common currency, citizenship, and foreign policy. This last[5] to 'include questions related to the security of the Union, including the eventual framing of a common defence policy, which might in time lead to a common defence',[6] for which task the WEU was designated executive agency. Yet signatories were further to 'respect the obligations... under the North Atlantic Treaty and be compatible with the common security and defence policy established within that framework'.[7] Meanwhile the WEU itself drew up an agenda at Maastricht in a Declaration of the same name which outlined cooperation with NATO, the Planning Cell, the Surveillance Centre and an eventual WEU armaments agency. WEU activity in the area of armaments was subsequently elaborated to absorb that of the IEPG[8] and the now inert NATO Eurogroup.[9] Although these measures have the ostensible effect (reinforced by the admission of new associate members to the WEU) of strengthening the link between NATO and WEU, and thus the North American connection in European security, the creation of a European pillar of NATO inevitably means a degree of polarisation within the Alliance. The details also chip away psychologically at the fence surrounding defence production under Article 223 of the EC's Rome Treaty, on top of the natural erosion it could be expected to suffer in the Single Market.

Three Key Players

I—Germany

At the highest political level, the Atlantic hinge came under strain during 1990-91 as a result of the chequered European response in the Gulf and the pre-Maastricht posturing. President Bush signalled his concern at the French-inspired drift in the Alliance at NATO's Rome summit shortly before the Maastricht IGC met. By then the Kohl-Mitterrand proposal for a European-only military force had sharpened the tone not only in Washington's relations with France, but with America's key partner on the European mainland: Germany.

Indeed in 1989, the *annus mirabilis* of Eastern European revolutions, Bush had been the only leader among Germany's major allies to acknowledge the end of her truncated post-war existence with so much as good grace. Germany's unity, after all, was professed NATO policy and Bush was not handicapped by the historical nervousness displayed by Mitterrand or Thatcher. For a short, promising period it seemed that a new central partnership was being forged between the natural leader of the democratic community of nations and Western Europe's key economy. *Die Welt* extolled the United States as Germany's 'most important and reliable international advocate'.[10] But the US-German courtship foundered in the coming year as Bonn hid behind the constitutional fiction ruling out deployment outside

NATO, Kohl pledging instead ever more apologetic subsidies and materiel support. The helpless government was further embarrassed by opposition-led soul-searching as to whether Germans should really be supporting their Turkish allies. As the United States defended common interests in the Gulf, Germany appeared to agonise about low-risk assistance, timid and embarrassingly conspicuous. A similar 'European' fecklessness was to be contrasted with the apparent American wish to confront Serb aggression in Bosnia, which Germany has been accused of having triggered while yet declining to move to resist by force. Catching the national mood following DESERT STORM and facing re-election, Bush promised that his country would not accept the role of the world's policeman; his successor is overseeing a process of disengagement which is still to run its course.

An indication of how specific US perceptions of Germany were shifting was given by State Department leaks in early 1992, charging Bonn with 'assertiveness'[11] even though American policy suffered less from Germany's alleged diplomatic bullying than did the European neighbours. It was pointed out how Germany had frog-marched partners into recognition of Croatia and Slovenia within days of signing up for a European Common Foreign and Security Policy at Maastricht. The same *Mark* colonising the East was also bringing the other compliant economies of the European Monetary System (EMS) under increasing strain, and Kohl was demanding that the future European Central Bank be sited in Germany. Kohl had also overturned the careful balance of the European Parliament by demanding 18 more seats for the new *Länder*, and rumours were growing of a bid for a place on the UN Security Council. This high-level reappraisal coincided, at least in the editorial pages of the leading newspapers, with an appalled fascination with the ugly reappearance of racist violence. Anti-semitism touches a raw nerve in the United States, particularly when found in the birthplace of Nazism. Combined with the growing awareness of the country's flagging economic competitiveness, it was clear that America's model

European was losing her lustre. Nonetheless, as subsequent events have shown, engagement with Germany remains important even if expectations have now fallen.

II—Britain

By way of contrast, Britain could not have proved a more loyal supporter of the United States in the Gulf, as indeed she had proved over divisive NATO issues such as INF deployment and Lance modernisation where Germany had seemed to waver. The United Kingdom has indeed been a privileged partner within NATO in the fields of cooperation in intelligence matters and nuclear missile technology, but she has accepted the role of junior partner in this relationship while her own star has continued to dim. Britain has thus defended her own interests through a tenacious Atlanticism and resistance to any rearrangement of NATO which could sap America's commitment to Europe. In this she has not been alone: setting aside smaller nations like the Dutch and Portuguese, the Germans too have maintained a consistently strong and positive American link as one of the two key co-ordinates of foreign policy. What distinguishes the British, as a former great power, is the manner in which they have stressed the importance of Europe's open transatlantic link in contrast to the maximalist, federal European project. Since the latter was sponsored by both France and Germany, Britain's historical claim to hold the balance of power was invalidated, and the country has become marginalised. When able to do so, Britain has used her highly-proficient military as an time-honoured instrument of national prestige and international influence—'Punching above one's weight', as the Foreign Secretary put it—but her inability to offer a larger force in Bosnia has shown up the limitations of this policy.

As a result, although Britain's resistance to the accelerated deepening of the European Community has been fought in the cause of the wider Atlantic Alliance, Washington's verdict that (for all the drawbacks) backing the European majority view serves America's interests left the British hanging. An ex-State Department commentator observed that '[Britain] should realise that nothing is more demeaning than to be supplicant to a master who no longer has use for you'.[12] Meanwhile, with the departure of the Reagan-Bush *ancien regime* in Washington, the Special Relationship has shrunk to an association between military establishments. In generation, in his vision of statesmanship, Clinton differs in every respect from his recent predecessors. It might be said that, as a former Rhodes Scholar at Oxford, he has direct experience of the linguistic and cultural ties that make personal communication between Britons and Americans so natural and effortless (factors which partly explain the UK's position as the leading European investor in North America) but he has done nothing to emphasise this. Predictably asked by reporters about the state of the Special Relationship at the time of his first meeting with John Major, America's new president did not even abide by the conventions, replying that 'we have many special relationships'. Worst of all, in foremost matters of US-European discussion such as trade disputes or common action in Bosnia, for all her good intentions, Britain is no longer being treated as a constructive supporter but as part of the problem.

III—France

For France, creating precisely this polarisation has been a deliberate policy aim. In direct contrast to that of Britain, France's answer to her post-War decline was to define herself *against* the United States from within the safety of the NATO umbrella. Since policy has been formed on the basis that what is good for France is good for Europe, the post-Cold War Europe of the Maastricht Union was to be a politico-economic superpower, unambiguously empowered to face down North America and the Far East.

Alone in Western Europe, France's president retains an executive role, above all in foreign policy; in this Mitterrand faithfully pursued de Gaulle's line to its logical conclusion. He redoubled efforts in

1991-92 to minimise the US and NATO roles in Europe, reviving his European-only 'Confederation' project, promoting a hard security role for the Conference on Security and Cooperation in Europe (CSCE), pointedly refusing to participate in the NATO-inspired North Atlantic Cooperation Council (NACC) then forming, and for a period during 1992 blocking NATO's attempts to adapt to a UN peacekeeping role. Between October 1991 and April 1992, while Europe wrestled with the continent's first war in 46 years, he drew Kohl into the bid to arm Europe with an autonomous defence capability through the European Corps, a dynamic pursued elsewhere through maritime cooperation with France's Mediterranean neighbours and more importantly through persistent moves to seduce Britain into a closer nuclear partnership. Even while France has been observed to soften her obstructive position—accepting the principle of NATO's peacekeeping function and the Alliance's authority over the European Corps in wartime—she has continued to contest the *status quo*, as in issues of Command and Control in Bosnia. Yet to conclude that this amounts to America's deliberate step-by-step expulsion from Europe is to overstate the case.

French protests that they do not seek to drive the Americans out of Europe should be believed, for America represents France's final insurance if Germany breaks out of her embrace. The fear of this was apparent in government appeals to the panic vote during the Maastricht referendum campaign,[13] and indeed General Galvin, as Supreme Allied Commander Europe (SACEUR), advanced this very reason for America's continued and reassuring military presence.[14] Mitterand has believed, with every reason to hope it, that NATO's process of 'Europeanisation' goes further than has yet been admitted. Meanwhile France's military potential gives this difficult partner the means to underline her importance by cooperating conditionally with the US, in the Gulf, Kurdistan, Somalia, Cambodia and the former Yugoslavia. Having accepted peacekeeping at NATO's 1992 Oslo ministerial as a legitimate Alliance activity while stipulating that member states are not obliged to participate, as a permanent member on the Security Council, France herself (like Britain) can rarely afford to decline to intervene. This policy has a price, but high stakes have characterised French poker style throughout.

Troops and diplomats are not, however, the only instruments at France's disposal. Following the 1993 elections in France, the CIA released information relating to previous campaigns of alleged industrial espionage since 1990,[15] putting the new government on guard. More openly, France's state-owned industries have been employed as pieces in the Transatlantic chess-game, giving rise to Airbus, Eurocopter and Euromissile. While France declined to participate in the SDI-derived GPALS programme on the grounds of her opposition to the militarisation of space,[16] she herself has been the prime mover for European space programmes (including military surveillance) and had a commercial interest in GPALS research through EUROSAM. Neither is the commercial challenge being made only through leading technologies with a defence application, as is demonstrated by disagreements within the Community and the GATT over agricultural subsidies—a clear instance of France staking the pitch for a trade war. In 1992 all French political parties were united in their opposition to moves to facilitate agreement in the Uruguay Round, denouncing the necessary concessions suffered in reform of the Community's Common Agricultural Policy simultaneously as vandalism of France's cultural heritage and as a craven submission by her Community partners to the rapacious demands of Anglo-Saxon ultra-liberalism[17]. Having thus stoked the anti-American farm protests of 1992, Bérégovoy's Socialist government was then swept away in a tide of inchoate discontent at CAP reforms, the single market, immigration, unemployment and the personal unpopularity of the president and his ministers. Since recent history weighs heavily on France, current and future governments are unlikely to escape this legacy. Having promoted (formerly with German acquiescence) a vision of Europe with herself at its centre and by persisting with an apparently visceral anti-Americanism, France bears conspicuous responsibility for the deepening transatlantic rift.

The Hour Of Europe

This rapid and alarming deterioration in transatlantic relations can in retrospect be seen to have passed a milestone in the summer of 1991, when Yugoslavia finally imploded. With the brief Slovene war over, the Community attempted to retrieve its unity and to assert the EC's collective statesmanship. In his capacity as president of the EC, Luxembourg's prime minister Jacques Poos grandiloquently declared that the hour of Europe had come.[18] What followed was a toothless mediation effort, later passed up to the UN, the ineffectual despatch of white-coated observers, and from late July 1991, endless circular discussions on the possibility of serious military intervention, feeding into the pre-Maastricht tug-of-war over responsibility for European defence.

Washington, similarly caught out by events but an ocean away, held back to leave the Europeans a clear field. By mid-1992, Sarajevo's worsening plight was making such detachment increasingly difficult, but after prevaricating in identical fashion to avoid being sucked in, both Europeans and Americans, WEU and NATO vied with one another to intervene while yet avoiding the risk of entanglement. Finally Bush settled on more decisive activity over Kurdistan and in Somalia, while Bill Clinton was elected that year on a raft of America-first commitments which also included a pledge to 'do something for Bosnia'.

On top of the patchy response of the European allies to Saddam in 1990-91, America found confirmation in Yugoslavia that 'Europe', as used by its propagandists, did not yet exist. As a mental shorthand for a rich economic space characterised by fractious and unreliable politicians, the term came increasingly to be used by the political classes and opinion leaders in sorrow or condescension; for their taxpayers and readers with preoccupations closer to home, it played on folk-memories of an Old World in perennial thrall to disunity, war, and race persecution. Decreeing a Common Foreign and Security Policy at Maastricht had not endowed the EC's 340 million citizens with sufficient common will to extinguish a war on their own borders. The upheavals in the currency markets of September 1992 and the forced ejection of two G7 currencies further undermined Maastricht's credibility, Indeed, considering the no-vote in the June 1992 Danish referendum and the 51.4 per cent Yes of the French voters in September, half of the Community's citizens had no appetite for such a union anyway. The US had adapted to the Maastricht vision to the point of chiding the British for dragging their feet, yet by late 1992 the promises were looking worn. And still the Single Market which had earlier excited American hopes in Europe's development was not officially open.

1. US Department of State's Rozanne Ridgway, European Wireless File 18 June 1989.
2. Art.30, para.6(a).
3. Art. 1.
4. Title III.
5. Title V, Art.J.
6. *Ibid*, para 1.
7. *Ibid*. para 4.
8. NATO Oslo meeting, March 1992.
9. NATO Ministerial, 25 May 1992.
10. 13 December 1989.
11. See *The Times*, 9 January 1992; *The International Herald Tribune*, 11 January 1992; Hanns W Maull, 'Assertive Germany: Cause for Concern?' in *The International Herald Tribune*, 17 January 1992.
12. John W Holmes, 'A New Special Relationship for Britain' in *The International Herald Tribune*, 2 February 1993.
13. See *The Independent*, 15 July 1992; *The Financial Times*, 28 August 1992; Geoffrey Hodgson, 'Side by Side, Masters of Disaster' in *The Independent*, 17 September 1992.
14. *The Times*, 16 October 1991.

15. *The International Herald Tribune*, 28 April 1993.

16. For a French establishment articulation of the objections, read Dominique David, 'Defense antimissiles: les enjeux' in *Libération*, 13 April 1992.

17. Jacques Chirac described this as 'an agricultural Munich'; see *The Independent*, 23 November 1992; the Socialist government's industry minister, Dominique Strauss-Kahn, even spoke of a domineering US wielding a 'food weapon over the rest of the planet', *The Independent*, 21 November 1992.

18. *The Financial Times*, 29 June 1991.

THE GRAND DESIGN

1 January 1993 was to have been the defining moment for the member states of the European Community, symbolising a new dawn of European trade cooperation. In the euphoria, John Major proclaimed that the EC's huge internal market would be open for business.[1]

The progression towards the creation of the European internal market had its origins in 1987 with the adoption of the Single European Act (SEA). It was hoped by its formulators that the adoption of the SEA would stimulate economic growth and efficiency for all EC member states by creating a single market of comparable size and operation to that of the United States or Japan. The prospect of a barrierless, homogeneous internal European market created a plethora of expectations both within and outside the Community. It was defined as 'an area without internal frontiers in which the free movement of goods, persons, services and capital is assured'.[2] With this, the EC appeared poised to become a powerful economic force in the global marketplace. Yet to some outside observers, the reduction of internal barriers raised the spectre of a 'fortress Europe', which while liberalising trade within, would remain protectionist to outsiders. Indeed this has been the fear of much of American industry. US policy, on the other hand, has stressed the advantages of a mechanism which would force Europe into playing a much more positive role in the transatlantic alliance.

Progress towards this goal since the beginning of 1993 has been encouraging, for out of the 282 measures in the single-market programme, 262 had been agreed within the first quarter of the year, and the vast majority of the 241 required to be implemented had been incorporated into national law by member states.[3] However these statistics fail to convey the whole picture. While such single market legislation caused little actual pain, it was easy in principle to remove customs checks for road haulers and restrictions on capital, and to

17

advertise public contracts. Once these directives began to affect local employment, however, certain member states balked. This is why telecommunications companies have been given until 1998 to open up their operations to outside competitors and energy liberalisation has been delayed.[4] Furthermore, no one doubts that although the public procurement market may be open to outside bidders, they compete at a disadvantage. For this reason, public agencies in most member states will continue to use fleets of vehicles, computers, railway stock manufactured (if appropriate) in that same country. What is more, national technical standards continue to be deployed in defence of protected national markets as the risible debate over hops and water in beer indicates, although the Commission has had some success in eroding these barriers. In areas such as steel production, where significant overcapacity exists within the Community, governments risk paying a heavy electoral price when slimming down to conform to diminished opportunities.

The problem of national, as opposed to 'European' interests, is further entrenched by the structure of the Community's legislative bodies. Following Maastricht, the powers of the Commission's bureaucrats remain overshadowed by those of national governments or similar appointees (such as Germany's independent central bankers), thus perpetuating national rather than 'European' public attitudes—particularly in countries with strong national identities. For this reason, legislation aimed at opening up high-technology industries such as telecommunications has been delayed, in this instance through the vigorous defence of national monopolies by France, Belgium and Germany.[5] The reason for this is that 'high technology industries remain vital to a nation's well-being, not only because of their immeasurable contributions to exports, high wage and high skill jobs, productivity, and R&D, but also because of their immeasurable contributions to the nation's technological capabilities'.[6] If member states of the Community allowed these sectors to become open to a deregulated free market, many would soon find their high technology industries uncompetitive, leading to bankruptcy or takeover by external competitors. Despite the single market legislation, much high technology industry thus remains the domain of individual member states, protected from competition both inside and outside the Community.

Recession has also constrained member states' willingness to cooperate for the wider European good and is, in the process, causing a reappraisal of the whole concept of 'Europe' as envisaged by the Maastricht treaty. The clearest indication of how the mood is changing is provided by the fortunes of the European Monetary System (EMS), established in 1979 and subsequently given a deadline for monetary union (EMU) in the Treaty. Indeed, the Treaty effectively decreed that the national economies in the EC would converge to enable this. While economies were expanding, the ERM passed on Germany's low inflation discipline throughout the Community. However, when the costs of reunification, coupled with recession, hit the German economy, the other EMS currencies were forced to endure high interest rates at a time when the very opposite was needed in order to spur recovery. The Deutschmark's partners could not indefinitely maintain the high interest rates demanded by the *Bundesbank* with economies already suffering from deepening recession. In September 1992 the first upheaval forced Sterling and the Lira out, and the Peseta, Escudo and Irish Punt to devalue. By summer 1993, the point had been reached where either the besieged French Franc (and Danish Krone) were ready to be ejected through the currency grid's floor, or the Mark would need to be freed upwards. Under the latter proposal, the Mark would have been followed by the Guilder, then the Belgo-Luxembourg Franc. This 'Jeudi Noir' effectively put an end to the EMS altogether, since a grid which allows fluctuations between currencies of up to 30 per cent imposes no discipline worthy of the name. The currency crisis of August 1993 served as a warning against this prescriptive model of building the Community. Regardless of the rhetoric of Germany's political leadership endorsing a 'European' ideal, the bankers of Frankfurt still give priority to their national responsibilities. Germany occupies the central and pivotal position in the EC's development, but officials elsewhere are no less preoccupied with national priorities than is the *Bundesbank* with the challenges of reunification. Recession, exacerbated by Germany's own economic woes, has caused Europe to falter just when it seemed poised to realise its potential.

Industrial Policy: A Lack Of Consensus

The main aim of Single Market legislation was to rationalise industry along European lines by eliminating unproductive firms and reducing the inherent overcapacity in many sectors. Industrial restructuring enables national industries to merge with their rivals to create industries on a European scale, while firms which had a comparative advantage over others are able to further consolidate their market position *vis-à-vis* their main rivals. As a result, EC firms would then be able to take advantage of the resulting economies of scale and compete more effectively in global markets. No longer would the EC be riddled with policies that maintained lame duck industries.

Although this grand design was good in theory, in reality, economic rationales have become subordinate to the will of national governments. The Common Agricultural Policy (CAP) is a case in point. Reform of the CAP has been actively pursued by member states but intransigence, notably by France, has reduced the impact of the Community's attempts to rationalise the agricultural industry. Now farmers are paid money to leave some fields fallow and, as a result, European consumers still have to pay for overpriced produce. Indeed, in a year when incomes will rise by no more than two per cent, many farmers can expect theirs to rise by at least 20 per cent.[7] One reason for the slowness of reforms is that the French farmers can generate huge domestic support for their cause as their predicament matches the country's protectionist tendencies. Indeed, Prime Minister Eduard Balladur was elected on a wave of popular support of protectionism for French industry and in particular his vocal opposition for agricultural reforms. In effect he is now hostage to this commitment and has, as a result, said he will not support the Blair House agricultural agreement (the US/EC deal aimed at lower tariffs in order to get an agricultural deal in GATT) which he believes will benefit the Americans over the French. Balladur has continued with his predecessor's trade rhetoric, threatening to undermine the Community and not follow through with changes he fears will create a wave of cheap foreign imports.

The very fact that Blair House is threatened illustrates the fact that 'European' interests are hostage to the individual will of national governments and that the EC's industrial policy is undermined by these actions. France has a history of labour militancy, particularly in such areas as agriculture and fisheries, and the violent protest that followed the November 1992 agreement to modify the CAP was clearly predictable. However, it rebounded on Pierre Bérègovoy's Socialists, contributing to their fall from grace the following April. Balladur faces the continuing challenge of trying to reconcile the national interest (widely perceived as that of the uneconomical peasantry) but avoid a Franco-German breakdown over the GATT negotiations.

The difficulties the Community faces in trying to overcome government hostility has been further demonstrated in the steel industry, which is similarly bloated. Although Italy, Spain and Germany have accepted the need to rationalise they cannot agree on how much capacity each should shed. This lack of consensus has forced Karel van Miert, the competition commissioner, to emphasise the threat this poses to the credibility of the Commission, while at the same time putting at risk its ability to solve state-aid cases. If left unresolved, this could allow states to adopt go-it-alone and beggar-thy-neighbour policies.[8] The Commission faces the problem inherent in any restructuring of industry of how to force through policies that will invariably lead to national job losses (in the case of steel, these would be around 100 000). Whether in the steel industry or farm production, rationalisation entails unemployment in the areas affected; with unemployment across the Community already nearing 12 per cent of the labour force,[9] this is a highly-charged issue for national governments. The government in Bonn feels compelled to shore up some of the remaining heavy industry in the eastern *Länder*,[10] while its counterpart in Madrid has encountered fierce domestic opposition to further mill closures. Shooting lame ducks therefore carries a social and political price at home.

These conflicting interests within the EC have, however, even graver implications as they threaten to undermine the commercial unity of the Community. Without this, the EC will find it increasingly difficult to meet the new hard line trade threat from the United States as expressed by its trade negotiator Mickey Kantor. Furthermore, if the Community fails to find common ground over issues such as agriculture and steel then it is unlikely that high technology sectors can be rationalised. This will enable the US and Japan to target these sectors with little fear of inviting trade retaliation from a unified and strong European Community.

Towards Uncompetitiveness

The fact that overcoming national interests still dominate much of the Community's work can be seen in its attitudes to the process of rationalisation. For if full integration were to occur then industry would relocate at the cheaper labour source as this is the essence of the free market and economies of scale. Fear of this industrial relocation as a result of the North American Free Trade Agreement (NAFTA) has led to the rhetoric, expressed by the independent presidential candidate Ross Perot, of jobs being 'sucked' to Mexico due to its cheap labour costs. The Community, in an attempt to eliminate both this structural problem and 'social dumping' (the deliberate act of governments reducing labour costs in an effort to attract inward investment), sought to harmonise social costs throughout the Community by means of the EC's Social Chapter.

In practice, this threatens to pass on high unit labour costs throughout the Community. It already costs US$26.00 an hour for a German worker to fit wheels on a VW. A Pole or a Chinese will do the same job for less than a dollar an hour.[11] Furthermore, while West German wage rates are already the highest in the world, those of most other European countries are also more expensive than either America or Japan. In essence, in attempting to create a uniform labour standard, the Community has exacerbated its uncompetitiveness in the global marketplace.

The UK has stayed aloof from the Social Chapter since Maastricht, arguing that workers are already sufficiently protected by national legislation and that such measures are anti-competitive in practice. When the Hoover company decided to take advantage of the single market and relocate from France to Scotland, it seemed to herald the beginnings of industrial restructuring based on the costs of labour. Hoover had merely capitalised on the opportunities presented by the single market, consolidating production at one European plant so as to generate economies of scale of 25 per cent and take advantage of the UK's low non-wage labour costs. The outcry by the French government against 'social dumping' in February 1993 was further evidence of the obstacles that exist to prevent Europe restructuring along purely economic lines. Attempts to prevent industries from locating at the cheapest labour source therefore emphasise Europe's division over how far to progress with a programme of industrial rationalisation.

Further Divisions In The High-Technology Sector

In some cases cross-border mergers have occurred and overcome the inherent political opposition to any industrial changes. However, this is usually due to the fact that governments find themselves in a 'no-win' situation. Either they accept some job losses but remain in production, or resign themselves to the possible loss of an industrial capacity in that particular sector. The recent acquisition of 51 per cent of Dutch aircraft manufacturer Fokker by Deutsche Aerospace (DASA) is one such case. Although DASA gains a controlling interest in Fokker through collaborative agreements, the coalition preserves Fokker's role in an increasingly competitive industry and secures long-term financing for future programmes.[12] Fokker's decision has therefore been viewed as a necessary precaution against further 'poaching' from other competitors, who may not have offered the same restructuring package.[13] In this event Fokker could have lost the ability to participate in future research and development. In instances such as this, restructuring is coming about as a policy of last resort. Strategic mergers need to become more common if EC firms are going to rationalise their production and operational costs. Only by encouraging this will Europe be able to compete on an equal footing with its many trade competitors. But in high technology sectors, government intervention continues to be the norm regardless of SEA legislation.

Even where it appears that Europe is beginning to restructure along economic lines, the alliance or collaborative project is underpinned by government involvement. In such cases it is more often than not governments that determine which partner develops a particular element of the project. On occasion, market leaders in given technologies have therefore found suitable work allocated to other-nation industrial partners. Most EC-backed joint ventures start with open-ended administrative structures for cooperation rather than with clearly-defined projects enjoying support from national industries.[14] Sometimes this has been because compromise remained out of reach once the project moved to the definition stage—the history of military collaboration is littered with such episodes, of which the French departure from the European Fighter Aircraft (EFA) consortium is perhaps the most conspicuous European example. At other times, the urge to promote an EC standard has led to blindness concerning developments in the outside market. In their determination to develop a European system, the EC pumped some £464m of taxpayers' money[15] into a High-Definition Television project, only to discover years later that the system being developed was likely to be obsolete before its launch, as it was incompatible with other systems already under development elsewhere.

Airbus Industrie, on the other hand, is one example of a successful European collaborative project. The Airbus consortium members, (France, Germany, Spain and the UK) have managed to capture a large market share of the commercial aerospace market and are now second only to Boeing in aircraft sales. Airbus' rise is testimony to the fact that, once clear parameters are set, European high technology collaborative projects can be competitive.

Yet although Airbus is portrayed as the model collaborative project, in reality it again demonstrates the protectionist instincts of EC member states. The Airbus consortium is not a European company with fully integrated research, development and production facilities, but is merely a marketing organisation. Construction of the finished product may take place within one state but the actual construction of the parts takes place in numerous sites around Europe: freight doors in Baden-Württemburg, tail parts in Seville, forward fuselage parts in St. Mazaire and Naples, and wing assembly in Chester. In many cases, workers divide their time between Airbus jobs and tasks for the company that employs them.[16] In reality therefore Airbus is not a collaborative project in the true sense and suffers from duplication and unnecessary assembly costs because of its decentralised structure. The scope this structure offers for Airbus to become the subject of a tug-of-love between its partners is illustrated by the manner in which Deutsche Airbus had to give up work on the interior of the A340 model in order to wrest assembly work on the A321 from Toulouse. Issues such as these and the political rivalries they represent colour industrial policy, with even perceived European successes being riddled with government intervention.

Airbus was in fact created by governments with the express aim of breaking into the aerospace industry, hitherto dominated by the

Americans; it was therefore above all a politically rather than economically motivated project. Indeed its political profile has remained high. Throughout his election campaign Bill Clinton focused on the consortium, claiming that it received over $13.5bn in trade-distorting government subsidies and, for this reason, would increase its market share at the expense of its American competitors. The Europeans retorted that the US aerospace industry benefits from indirect subsidies through military contracts. The figure claimed by Washington is hard to verify, but it is obvious that Airbus is operating within a privileged market in Europe. The national carriers of most EC countries operate Airbuses as the mainstay of their fleets (unlike British Airways and KLM, which stand apart as being in private ownership). This form of national preference has been referred to above. Yet the relationship is closer still, since at least in the case of Aerospatiale (holder of 37.9 per cent of the consortium), a state-owned manufacturer is supplying the state-owned airline. The pattern produced thus mimics that found traditionally in the manufacture and procurement of defence equipment. It is particularly striking that exact parallel partnerships exist to produce and supply missiles and military helicopters, built around the same core DASA-Aerospatiale axis—itself a mirror of the Community's central political partnership. Aerospatiale is now due to be privatised as part of the Balladur government's package announced in mid-1993; having been used to establish a European industrial platform, it will be freer to pursue other and more flexible partnerships in future, although the date for this measure remains some way off.

The hindrances to the unfettered operation of the Single Market are thus twofold. On the one hand, traditional industries which exist in overabundance are being defended by electorally-minded governments in sustained rearguard actions. On the other hand, meanwhile, key technologies with a narrower base are still being protectively nurtured—even by ostensibly *laissez-faire* governments— with an eye to the future. A true single market is therefore painfully evolving. Yet while national governments place national interests before Community interests, the EC's collective authority will remain weakened. And this, paradoxically, threatens those very high-technology sectors that member states are attempting to protect.

1. *The International Herald Tribune*, 28 December 1992.
2. See Single European Act and Final Act, Luxembourg, 17 February 1986 and The Hague, 28 February 1986.
3. *The Economist*, 10 April 1993.
4. *The Economist*, 3 July 1993.
5. *The Financial Times*, 29 April 1993.
6. 'Who's Bashing Whom? Trade Conflict in High Technology Industries', L. Tyson, Institute for International Economics, Washington D.C., 1992. p.42.
7. *The Guardian*, 7 August 1993.
8. *The Independent*, 8 July 1993.
9. Morgan Stanley International Investment Research, 15 June 1993.
10. *The Wall Street Journal Europe*, 29 April 1993.
11. Morgan Stanley, International Investment Research 15 June 1993.
12. *Aviation Week and Space Technology*, 30 November 1992.
13. *The Financial Times*, 2 November 1992.
14. *The International Herald Tribune*, 10 September 1992.
15. *The Independent*, 17 May 1993.
16. *The Wall Street Journal Europe*, 3 March 1993.

Regardless of the lack of political will to integrate some industrial sectors, the member states are nevertheless becoming more dependent on each other as economic forces drive industrial restructuring. Indeed, from 1961 to 1991, EC member states increased their exports between themselves from a total of 43 per cent to 62 per cent, while intra-EC trade as a proportion of the global figure rose over the same period from 15 per cent to 24 per cent.[1] Therefore the trading nations of the EC may be tempted to erect barriers against outsiders, but they cannot afford to be seen to do so between one another. The development of NAFTA, and the proposed formation of an Asian economic bloc are further indication that these alliances are growing as states begin to see the benefits of predictable and stable trading relationships. Trade blocs are therefore becoming official policy; inevitably bloc-to-bloc politics will hamper the creation of a multilateral trade system. The fact that trade blocs tend to perpetuate uncompetitiveness by reducing competition has tended to be overlooked. Instead they are viewed as alliances to counter discriminatory measures aimed at preventing open competition in the global market.

America has begun to adjust to these new realities and in the process has realised how much it lags behind others in many high technology sectors. Indeed, many Japanese civilian firms are often now more adept in some technologies, such as electronics, infra-red sensors and optics, and ceramics than are their US military counterparts. The main American argument is that once a high technology industry has gained a comparative advantage over others, it would then be able to maintain this edge by economies of scale and product sophistication, thereby excluding entry to others. This so-called 'first mover advantage' means that established firms could in theory maintain their market position indefinitely.[2] The new administration has therefore started to target foreign strategic high-technology sectors in order to prise them open for competition. This has been evident in the posture of trade negotiator Mickey Kantor and is implicit in the assumption that as the US divests itself of its Cold War burden such policies would become the order of the day.

Throughout his election campaign of 1992, Clinton reiterated his commitment to reducing the US budget and trade deficits and hauling the domestic economy out of its slump. To achieve this, Clinton believes that America must create 'free' and 'fair' trade, performed on 'level playing fields'. Subsidies must be reduced and protectionist mechanisms, such as tariffs, must be abolished. On the one hand, the administration claims that it remains committed to free trade:

> 'The fact is, despite rumours to the contrary, this administration is not a protectionist administration. We believe in free trade. We also believe in open markets...'[3]

In reality, trade is rarely wholly free or fair; playing fields are seldom quite level. Yet Washington continues to complain of disadvantages *vis-à-vis* its main competitors:

> '... We also believe in being treated fairly. We also believe that if companies from foreign countries have access to our markets, we ought to have access to theirs.'[4]

After an uncertain opening, the new policy has become apparent: the US has played the tough negotiator in the belief that it is far wiser to be feared rather than loved. Compared even to the position adopted by the Bush presidency, Clintons' Administration has been actively promoting a tougher trade stance. Indeed, Kantor rhetoric that the US will 'compete, not retreat',[5] has raised the spectre of trade wars not seen since 'those Smoot-Hawley days' of the 1930s.

Calls during his election campaign by Clinton and Laura d'Andrea Tyson, subsequently his chief economic advisor, for the reintroduction of the 'Super 301' legislation (part of section 301 of the 1974 Trade

Act) indicate how far along this road America may travel. The aim of 'Super 301' is directly to target a specific industrial sector deemed to have an unfair advantage over the US in an effort to open the market to US firms. 'Super 301' is an obvious protectionist tool intended to raise the stakes in negotiations over market access. The EC, however, must share blame for having cushioned high-technology industries from the full impact of the single market. Although the EC continues to put up a united negotiating position, its ability to maintain this has been brought into question. Deutsche Telekom, for instance, has been successfully 'seduced' by the US in favour of a bilateral deal with America in defiance of Community solidarity and, indeed, of the Rome Treaty.[6] This action coincided with German outrage over the anti-competitive preference given by Britain and France to bananas from their former colonies. These may prove to have been only the beginning.

The Gatt Forum

The Uruguay Round of the General Agreement on Tariff and Trade has become the focal point for the new fears and aspirations. A successful GATT deal would eliminate tensions in many sensitive trading areas such as telecommunications and agriculture, and open up previously closed, tariffed and subsidised industries to outside competitors by providing a forum for the long-term settlement of disputes. George Bush's much proclaimed 'New World Order' was based on the notion of greater cooperation between states, especially through liberalised trade relations. And while the 'fast-track' deadline (a clause allowing the president to negotiate treaties without pork-barrel congressional meddling) still stands, it offers a viable time limit for agreements. Despite a clear deadline of 15 December 1993, each side remains reluctant to compromise. Even negotiated agreements regarding agriculture have been met with disdain and hostility. As the GATT talks have dragged on, the changing political scene has affected the prospects for success. Rows have waged over public procurement policy and the EC's Article 29 'Buy European' policy.[7]

On this issue a solution was found, but the precipice draws nearer. Punitive steel tariffs imposed by the US on the EC constitute a further step towards the edge. The fact that it is acknowledged that sanctions damage the domestic economy as much as the sanctioned[8] indicates just how politicised the new relations have become.

The increasing acrimony surrounding trade is not confined to EC-US relations; it also dominates US-Japanese relations, with semi-conductors the main focus of attention. George Bush negotiated a 20 per cent share in the Japanese market for US semi-conductor firms. When this level seemed unlikely to be reached, the new administration demonstrated its confrontational attitude by threatening to implement sanctions. Conflict was avoided when it was announced that Japan had reached its agreed level, but in future the issue may not be resolved in time. This episode again reflects the irony of American trade policies and indeed the perils of managed trade. Now that American firms have achieved their 20 per cent stake in this high technology, highly profitable sector, they are unlikely to be granted the freedom to compete for a greater market share. The likelihood of brinkmanship—leading to open confrontation—increases when numerical thresholds are imposed to set strict limits. The net result of such political manoeuvrings has been that Japan has learned to say no to the US. Instead of a two-way fight, trade disputes have developed into a three-cornered tussle.

Growing from the semi-conductor episode is a realisation that for high-technology industries, bilateral deals are the most hopeful means to obtain market access. This could lead to a global trading system in which low-technology commodities are traded within the liberalised system foreseen within GATT, whereas high-technology components and systems are subject to bilateral agreements. Therefore, unless the member states of the Community are able to sink their differences in a cohesive group, the US and Japan will both chase bilateral access deals to its high technology sectors. Furthermore, and perhaps of greater concern, is that defence industries, given their high-technology know-how, will increasingly be the centre of attention.

1. *The Financial Times*, 23 August 1993.

2. S. Thomsen, 'Washington Thinking on Trade' *International Affairs*, Vol 69, No.3, p.538.

3. Ron Brown, US Secretary of Commerce, 10 March 1993.

4. *Ibid.*

5. Testimony before US Senate, 10 March 1993.

6. *The Financial Times*, 12 June 1993.

7. Article 29 states that non EC companies have to be more than 3 per cent cheaper in their tenders in order to win bids and must also have more than 50 per cent EC content.

8. *The Wall Street Journal Europe*, 24 June 1993.

DEFENCE INDUSTRIES AND THE SINGLE MARKET

I. The Nature of the Industry

The 1985 White Paper drawn up by Lord Cockfield and Jacques Delors, which provided the framework for the SEA, displayed a clear desire to see the defence industrial sector of the twelve come under the same single market legislation as all other sectors.[1] However, since 1985, progress towards this goal of an official and institutionalised single European arms market has been minimal. Indeed, a 1988 assertion by the EC Commission that the defence industries should become subject to the same tariff-less regime within the Community as other sectors met with little enthusiasm within the EC and with predictable hostility from the US.[2] The dysfunction between the European defence sector and the SEA as a whole centres upon the unique nature of the arms industry, since the production and sale of defence equipment does not lend itself to operation within the free market. National governments play a dominant role in directing what their national defence companies produce and to whom they sell the finished product.

Article 223 of the Treaty of Rome explicitly exempts defence equipment from the provisions of Community legislation, although use of this is a mask for protection of equipment 'not intended for specifically military purposes'. Since this was drafted, the dividing line between military and civil technologies has become increasingly blurred, with the result that economic reality is bringing European defence production within the jurisdiction of the SEA even in the absence of movement at a political level to this effect. Regardless of this lack of political will, companies within the West European defence industrial sector are already predominantly involved in non-defence related production. This trend of conversion from military production to civilian sectors is accelerating, in response to a reduction in demand for weapons procurement. Indeed out of the top fifteen worldwide defence firms, only six rely on defence for over half their yearly revenue.[3] Moreover, distinction between civilian and military industries is further eroded by the fact that many components have a 'dual-use'. Systems developed for civil purposes are now more often integrated into weapon systems than the other way around.[4] This question of dual-use equipment has been most clearly illustrated by the European technology and R&D initiatives such as ESPRIT and EUREKA in the civilian field and in the defence-oriented EUCLID programme. The much larger funding for the first two attracted the participation of companies such as Aerospatiale and British Aerospace and it is clear that some of the products of research carried out under these programmes will have a military application. Thus, the ostensibly clear distinction between defence and civil industries is in practice not all that simple.

Consequently the logic behind trying to regulate defence trade indirectly will become increasingly tenuous. Article 223 is fast becoming obsolete and will have little or no influence on defence firms in the future. Its decline, however, has important implications. For without it the regulation of defence industries will fall under the auspices of the EC Commission. Since dual technologies will permeate into wider industrial sectors, governments will be inclined to intervene with growing regularity in order to protect and maintain their technological advantage over others. The Commission will therefore be brought into further confrontation with member states as it attempts to open up these sectors to the workings of the single market.

II. Institutions

Despite occasional forays into the spheres of defence and security, Article 223 remains in place and the EC remains an institution without a formal mandate in these fields. This line is becoming indistinct—the Community's frustrated attempt to manage the crisis in the disintegrating Yugoslavia should be considered a bid to overcome it—and should the EC find itself with local responsibility following a Bosnian settlement, it will have been weakened further. Yet for the time being (indeed, for the foreseeable future), foreign policy remains the province not of the Community but of the national governments of the European Union, and the limitations this approach imposes are easily judged from the disharmony that has characterised intergovernmental relations over the Balkans. Where EC issues lead to technical questions of defence policy, the link binding it to the Western European Union (WEU) has been reinforced. The WEU has performed its first useful operational duties in Gulf waters in 1988-89 and 1990-91 and, more questionably, in enforcing the maritime blockade of the former Yugoslav republics. However, the WEU still occupies an unsatisfactory position, being left hanging as a 'bridge' between the EC and NATO, of which it is deemed to constitute the 'European Pillar'. What it is unable to do, just as the cart cannot pull the horse, is force its member states to align their security policies in advance; a shortcoming which mitigates claims to maturity prompted the creation of the European Corps, the Operational Planning Cell and the satellite data centre.

In two respects only can progress claim to have been made. In its function as Europe-in-NATO, non-EC members of the Alliance have been taken in with associate status. This in turn has enabled Europe's multitude of defence-industrial fora to be streamlined. Whereas the WEU itself hitherto acted as platform for defence-industrial discussion, now it is doing so under the banner of the Western European Armaments Group (WEAG), which became a sub-agency of the WEU when the latter absorbed the Independent European Programme Group (IEPG), simultaneously making redundant NATO's informal Eurogroup. Given that the Alliance also possesses a further instrument for industrial coordination in the form of the Conference of National Armaments Directors (CNAD), this rationalisation is clearly a forward step. In accordance with the WEU's role as a key element of the Maastricht Treaty's European Union, it has paid particular attention to those areas where Europe lacks indigenous capability, most notably strategic airlift and intelligence gathering. These areas have grown in importance as the West revises its strategic concerns after the Cold War. Yet the very same global transformation which brought about this reappraisal of the inter-relationships between organisations has also caused a change in perception of the defence industry. Now, more often than not, they are viewed as strategic assets as much for their value to national economies as for the purpose of the equipment they supply. This is because, while opportunities at home do still exist for new or replacement capabilities, the armed forces of the West are in contraction; as a result, few organisations remain exclusively or even predominantly active in defence activities alone.

III. Restructuring And Projects: The Progress So Far

Consequently, instead of allowing defence companies to become subject to the single market, European governments have encouraged restructuring along national rather than European lines. The absorption of Alenia SpA, Italy's largest aerospace and defence contractor by Finmeccanica in February 1993, coming on top of the acquisition of MBB by Daimler-Benz that created Deutsche Aerospace, and British Aerospace's acquisition of parts of Royal Ordnance, is the latest in the series. This development holds out the prospect of industry supplanting national diplomacy, with national players forming alliances even within a supposedly unifying market.

The end of the Cold War gave rise to fears that NATO's government would 'renationalise' defence; fears which have been borne out to some degree in the organisation of their armed forces, but to a greater extent in the case of these large industrial groupings. Yet this does not suggest that the pattern of collaboration that has become the norm is likely to be broken, since on cost grounds alone, this alternative scarcely exists. With the precedents of the Panavia Tornado and the Eurofighter 2000 behind them, the aerospace manufacturers of Britain and Germany are unlikely to strike out on solo ventures in the high-performance fighter class. Partnership offers a key to survival in such otherwise prohibitive fields. Having dropped out of the original Eurofighter consortium in order to pursue the Rafale programme, Dassault is highly unlikely to produce another aircraft unaided. In fact, conventional wisdom suggests an Anglo-French partnership for the next generation of fighters to be conceived, since the stakes for both these long-established producers by this time will be running equally high. (The experience of Jaguar shows that this is possible under less constrained circumstances despite the politicised character of French military-industrial efforts. It has already been observed how Paris has employed Aerospatiale as an agent of the state in partnerships with German industry.)The same can be said for armoured vehicles; there can be little chance of the UK developing a purely national Main Battle Tank by the time Challenger II is due for retirement. For related reasons, the Italians have devoted much of their even scarcer resources to the development of a native tank design, the C-1 Ariete, so as to have a card to play in future partnerships.

III High Stakes And Exports

It is thus apparent that EC governments will continue to play an active role in defence procurement policy. Defence contractors are still dependent on governments even when they have been given a free reign. This has been demonstrated by BAe's involvement in Europe's Future Large Aircraft (FLA) project. BAe has continued working with the consortium developing the FLA by funding itself in the project feasibility phase. However, the UK MoD has shown its clear preference for upgraded C-130J Hercules; BAe will be forced to withdraw after the feasibility phase even if the company is prepared to pay the British share of the project definition that follows. It would be highly risky to proceed to this without a firm government commitment to purchase the aircraft as this would leave British Aerospace reliant on exports. Furthermore, in collaborative projects of this kind, decisions are taken by governments not firms. It is therefore likely that BAe would receive few of the industrial spin-offs in the absence of a government backer demanding *juste retour.*

It is not difficult to see therefore why this restructuring is occurring along national lines. First is the role of such industries as a factor of national security, although as remarked above, in Western Europe in the 1990s this importance is more psychological than real. (The United Kingdom was dependent on material assistance from allies in both the Falklands and the Gulf Wars.) Second, as a result of this relationship, the defence contractor is dependent to a greater degree on a monopoly domestic purchaser; hence national requirements have increasingly dictated design as the manufacturing industries have contracted to field a narrower selection of products. This particularly applies to major, big-ticket weapons systems. Historically, French industry has stood apart from its western partners by producing defence material that gave greater emphasis to export criteria. By the 1980s, however, the differing national requirements (if not the operational doctrines) of Britain, France, Germany and the US could be seen to have converged to a remarkable degree at the heavier and more technologically advanced end of the scale. This in turn had narrowed the opportunities for exports to the fastest-developing markets, or those with natural resources to fund highly expensive purchases: the countries of the Middle and Far East. While NATO defence forces contract by an average of 25 per cent and are forced to eliminate surplus equipment stock, the difficulties of selling to near-saturated outside markets are compounded by the emergence of new indigenous manufacturers in Latin America and the Far East, and new competition from established industries in Eastern Europe and the former Soviet Union. The vital importance of defence exports to the balance of payments of the EC's leading economies is easily apparent:

World Ranking	Company	1992 Defence Revenue US$m
4	BAe plc.　UK	6,065
8	Thompson Group　FR	4,860
11	GEC　　　UK	4,088
13	Deutsche Aerospace　GER	3,912
14	Aerospatiale　FR	3,499
25	Dassault　FR	2,182
27	Alenia SpA　IT	1,959
32	Giat Industry　FR	1,577
36	Rolls Royce Plc.　UK	1,364
37	Alcatel Alsthom　FR	1,338
38	SNECMA　FR	1,300
43	Matra Hachette　FR	1,052

Out of the 46 firms whose 1992 defence revenue exceeded US$1,000m 33 were US owned and one was Japanese.[5]

The need to gain exports has meant that lobbying for sales for the time being still remains a national focus, with the sales pitch for the tri-national Tornado being seen in the UK as a purely British concern. Come the stage when, say, Germany is able to pursue foreign markets more aggressively, some form of agreed marketing practice will need to arise to avoid damaging in-fighting. Even short of a common export policy, this will exert influence on governments with vested interests to form broader fronts. Furthermore, as costs for the development of new systems increase exponentially this will act as a further driver towards European collaboration. Although defence industrial collaboration in Europe has enabled Europe to produce a range of products which individual European states would have found difficult to fund or to develop, the policy of *juste retour* has created additional costs, delays and disputes over specifications, work shares and project leadership which has caused some programmes to collapse with considerable political acrimony and resentment. Government intervention, however, is likely to decrease as economics force collaboration onto the European defence industries. It is already apparent that defence contractors are moving into closer cooperation with each other and overcoming many of the obstacles that the SEA imposes.

Due to the high-technology make-up of this sector, however, it will remain heavily protected which will inevitably bring it into contact with the new hard-line trade rhetoric of the United States. For collaboration not only enables Europe to compete more effectively in global terms, it also allows participating states to continue providing employment for high-technology sectors, thus maintaining their first-mover advantages.

V. Transatlantic Cooperation

The political and military link between Europe and the United States has been enhanced through a myriad of economic contacts: Europe remains a very important market for US products. Many US companies invested heavily in building up their R&D, production and distribution activities on the European continent, which therefore represents not only an important source for customers, but also a considerable investment. Technologically, the increase in the cost of modern R&D has forced both US and European governments and companies to pool their resources in order to stay abreast of the latest developments. Consequently, these trends have resulted in a marked degree of interdependence in many areas. Clearly such a complex relationship cannot be severed—even if one of the parties wished to—without a great deal of difficulty and defiance of economic, political and strategic imperatives.

However, the new hard-line trade policy adopted by the new Washington administration has sent signals to this effect. Although the transatlantic alliance has coped with diverging interests before, it did so because ultimately the Soviet threat overrode all other priorities. Now, more than ever before, the economy has begun to dominate the agenda and as a result high-technology industries such as defence have become the prime concern. America always had the better end of the so-called 'two-way' street in arms cooperation and as a result its defence industries were able to find ready partners in the Europeans. The European defence contractors are now no longer so willing to enter into partnerships with the Americans as US firms still demand to be the prime in any joint venture. This effectively protects the US defence industry from European competitors and also prevents, to a certain degree, any transfer of technology that would naturally emanate from collaboration. Furthermore the very fact that the US has only procured the T-45 Goshawk, the AV-8B Harrier, the 105mm light gun and the Beretta pistol shows how little access Europeans have had in the past. Moreover, the streamlining of the US defence industry through conversion and the bottom-up review would seem to point to the fact that the US market will become less, rather than more, open to foreign defence firms in the future as its 'leaner' firms will be competing for fewer contracts. In effect, both the Europeans and the United States are becoming more *laissez-faire* where their defence industries are concerned as economics drives restructuring. But because there are now fewer contracts to chase, they are being driven to restructure away from each other in order to consolidate their position in the domestic market. This paradoxically brings them into confrontation with each other as they compete for declining export contracts.

Indeed protectionist barriers have already begun to emerge on either side of the Atlantic as both sides complain of the lack of access to each other's markets.[6] This trend is, however, much more political in Europe as it centres around the need to create unity in the face of the strong US trade attitudes. Serge Dassault, chairman of Aerospatiale, has termed Italy's preference for leased fighter aircraft from the US as scandalous, and a decision which could break aerospace cooperation in Europe.[7] Moreover as European defence industries move into closer partnership it follows that any prospective EC members will be expected to buy European. It was perhaps because of this that the US has called for a GATT-style arms trade agreement aimed at boosting transatlantic cooperation.[8] The fact that no such agreement has yet been reached indicates that defence procurement cooperation is, if anything, already in decline. This may not yet be terminal, but as Europe restructures its defence around its own 'buy Europe' policy then American protectionist instincts, as expressed through their trade rhetoric, will come to the fore.

VI. Forward Projection—The US, The EC And The Transatlantic Alliance

In the past it has been accepted that the transatlantic alliance could be characterised into two distinct parts. The main dimension and its overriding *raison d'être* was security which consequently dominated its secondary economic role. Now, however, this logic has been questioned—'the battlefield has been changed from military confrontation to acute industrial competition'.[9] As this becomes an established fact within the alliance it has meant that high technology industries and their importance to the economy have come to dominate much of the transatlantic agenda.

Although NATO has been frequently dogged by disputes concerning weapon systems and the direction of US leadership, these disagreements were normally settled within the NATO framework. Today, lacking a mutual threat, the reasons for compromise are fast disappearing. Instead NATO is faced with diverging interests between members as each attempts to capitalise on the 'peace dividend' by concentrating more resources in the trade theatre. Following this reappraisal, America began to realise that it lags behind in many high-technology sectors. This realisation coincided with the election of Bill Clinton whose major campaign focus was the economy and who has subsequently formed an administration around the belief that the economy is a vital component of national security. Now, without the fear of upsetting allies, the US appears to have become a 'reluctant sheriff',[10] picking and choosing when and where to commit itself.

With this changed attitude Washington has begun to address the trade policies which are believed to discriminate against US firms and, as this policy has taken shape, it is clear that the transatlantic relationship has been affected. This changed focus can be most clearly seen in the way both the US and the EC have shifted inwards in an attempt to further capitalise on regional trading agreements. On the one hand America has begun looking towards the Pacific and its lucrative Asian markets and also towards its own North American trade bloc, while on the other, Europe has begun to formulate plans to enlarge the community and assess how it should progress in the future.

Regardless of this EC re-examination, it seems clear from the defence industries that, although not yet actively encouraged, integration and cooperation is occurring. Progression towards a form of common European defence procurement will, however, further exacerbate tensions with the United States as high-technology markets become increasingly protected. Washington is determined to be a tough trade negotiator and is being matched step-by-step by the EC, who desperately want to retain their technological advantage and privileged status. At present, the EC is being undermined by its lack of unity and as a result the US is targeting individuals in an attempt to gain access before full integration of the EC's high-technology industries reaches maturity. Once this trend becomes politically as well as economically motivated, then the restructuring of the defence side of high-technology that is being witnessed will be seen within the myriad high-tech sectors. Inevitably this will lead to divergence not convergence in the economic dimension of the transatlantic alliance. Although this may take some time, it is clear from the US rhetoric and the EC's progress (albeit slow) towards a single market that divergence has already begun to be the order of the day. Indeed this trend has also been reflected in the security sphere with the arrival of a more economically-conscious US President.

1. N. Colchester and D. Buchan, *Europe Relaunched: Truths and Illusions on the Way to 1992*, Economist Books/Hutchinson, London, 1990, p.120.
2. *Ibid.*, p.120.
3. *Defense News*, 19-25 July 1993.
4. M. Brzoska & P. Lock (Eds.), *Restructuring Of Arms Production in Western Europe*, Oxford University Press, 1992, p.5.
5. *Defense News*, 19 July 1993.
6. *Defense News*, 19 July 1993. p.42
7. *Ibid.*
8. *Defense News*, 19 October 1992. p.4.
9. 'From Geo-politics to Geo-economics', *National Interest*, Vol. 20, Summer 1990.
10. *The Economist*, 19 June 1993.

1993 AND ONWARDS

1993 makes a convenient turning point at which to consider the future direction of transatlantic relations, the beginning of that year being marked by a handover of power in Washington and the inauguration of Europe's fêted open market. The disappointments that attended each of these in the weeks that followed have been described in Chapter One, over Bosnia and in the opening shots of an undeclared trade war. White House watchers on both sides of the water were dismayed to see the incoming administration pursuing not simply policies which broke with recent practice, but gesture politics which were successively abandoned or modified away as major obstacles confronted them: Bosnia, homosexuals in the military, domestic health care. The picture thus emerged of a presidency in drift within months of taking office, with the lowest approval record for its length of tenure, stumbling even before key appointments in the State Department, Pentagon and other agencies had been made.

Elected to tackle America's ills at home—'the economy, stupid' as his election campaign expressed it—Clinton's diplomacy with the Europeans has been remote when set against the style of his predecessor's. His emissaries have shown contrasting characters. Whereas trade envoy Mickey Kantor has employed a hard-nosed take-it-or-leave-it line in pursuit of commercial interests, the dour, modest Secretary of State Warren Christopher has acted as a mere diplomatic relay, even though the new National Security Council reportedly put the Balkans at the top of the incoming president's foreign policy agenda.[1] Unlike previous administrations, the President's National Security Advisor himself has scarcely been seen. The blurred statesmanship that has resulted has drawn unhappy comparisons with the last Democrat administration of Jimmy Carter, with the disturbing difference that Alliance cohesion during the Cold War was maintained through the compelling external force of the Soviet threat. NATO even survived (indeed subdued) a state of near-war betwen Greece and Turkey because of this. This community of risk, uneven though it was, no longer obtains.

NATO Burnt In The Balkans

Bill Clinton had entered office encumbered by election pledges to do more both for Bosnia and for America at home. Since soundings of public opinion showed a convincing majority against intervention other than of the humanitarian kind being practised already by NATO allies,[2] the new president was obliged to tout stand-off responses, inappropriate and low-risk, which in any case foundered on the alarmed opposition of Europeans and Canadians and required the implausible assent of the UNSC. By the summer Clinton was already turning away from the Bosnia imbroglio, cloaking his incapacity to solve the issue once and for all in the supposed obstructiveness of allies who had themselves already failed.[3] The resonance of this version of events—in which the 'Europeans', or the disarmed Lord Owen, whose shaky plan for peace Clinton refused to endorse until it was already irrelevant, carry the can for America's equal irresolution—bodes ill for the future; it has been accepted not just by American politicians posturing at home,[4] but also habitual Transatlantic commentators.[5]

One of these, the *New York Times*' William Safire, who has lambasted Clinton for not enacting these incomplete policies[6] drew an acid comparison between Clinton's performance over Bosnia and his earlier firm decree of US$44bn at the 1993 Vancouver summit for the beleaguered Boris Yeltsin and Russia's reform process.[7] Since the money itself did not materialise with the same ease with which it was pledged, Clinton is open to accusations of summarily eliminating another election commitment, but the contrast is apt inasmuch as it suggests the limits of America's strategic involvement in Europe, even though the real challenges to European security arise in the lands between the EC and the Russian marches. 'Only the presence in Europe of Russia...', opined the retiring US ambassador to the EC, 'requires the indefinite maintenance of an Atlantic framework for the management of European security'.[8] It was not this spirit that kept the United States in Europe after 1945. Another American diplomat was heard to remark around the same time, *a propos* Clinton's backing for Bosnia's 'safe' areas, that Europe was witnessing the 'end of an era of American leadership'.[9] Finally, the same sentiments were expressed with a different nuance by a newcomer, Undersecretary of State Peter Tarnoff, that in cutting her coat to suit her cloth, America would pursue a role in future which 'may on occasion fall short of what some Americans would like and what others would hope for'.[10]

Tarnoff's assessment was quickly disowned by both Christopher and the White House, but the weight of evidence is compelling. US policy will shadow Moscow to the detriment of other regions, unless a new nuclear menace threatens to arise, as in the case of Ukraine. This conditional disengagement further undermines not only the tattered credibility of the CSCE, already discredited by the Yugoslav wars, but also that of NATO as it expands beyond its function as a mutual defence pact, since it remains dependent on American material contributions. (Indeed, the US shift cuts the ground from under the NACC altogether, depriving it of any attraction for countries such as Poland). Yet at the same time the WEU, because of France's renewed interest in NATO's structures, is deprived of its principal sponsor to the primary role in European defence. Such aspirations died anyway following the debacle in the former Yugoslavia. While the WEU is gathering assets, a revival in its fortunes can only come about when Germany overcomes her constitutional inhibitions to endow Europe with real military muscle. It is an open question whether greater coherence in Foreign and Security Policy will by then have been reached.

A Tangled Strategic Web

Britain and France have already been driven to renewed coordination by the force of events; these two Europeans are best equipped materially and psychologically for military intervention in the Middle East and the Balkans and may be drawn closer together by the need to pool aspects of nuclear policy. This prospect is brought perceptibly closer by Clinton's decision to extend the ban on nuclear testing (a facility of immeasurable value to the British). While the scaling down of both British and French nuclear armouries narrows the scope for such cooperation, such a tilt in Britain's strategic alignment will have far-reaching implications for the whole Atlantic Alliance. Moreover, once Germany has reconciled her population to the necessity of the military component of international power politics, Europe's strategic triangle will be complete. This transformation is already under way: Germany's tough regime on immigration in Central Europe illustrates how Bonn will move unilaterally unless NATO and the Community are more successful in internationalising her strategic concerns. Finally, Clinton's courtship of Germany as a 'partner in leadership' and support for a place for the country on the Security Council (along with Japan) adds further uncertainty to the pattern of future alliances and presages an uneasy defence by France and the UK of their post-War privileges.

Although the newly 'assertive' Germany was soon sidelined once the focus of Western diplomacy shifted to Bosnia because of her inability to commit forces, Germany is still left holding the key cards regarding the future of European security. She cannot but be simultaneously the key partner for the US, France, Britain, all the smaller immediate neighbours, Russia, Ukraine and Turkey. By 1995, the Community is also expected to have its membership and finances boosted by the accession of Austria and the Nordic states, swinging the EC centre of gravity north-east and away from the Rhine. An enlightened and vigorous government in Bonn (later Berlin) could pull the continent together despite the inevitable muttering; an indecisive and introvert one in thrall to public despondency will be its weak link. Chancellor Kohl is certain to be returned to office in 1994, if only because neither the opposition nor his own Christian Democrats will have been capable of putting forward another credible contender. The prospect is a dismal one; although the cautious, broad-front consensus politics of the Bonn tradition are unlikely to survive his eventual departure, Germany is in urgent need of a chancellor with the vitality and political courage to confront the challenges at home, and to lead the country to recognition of its international responsibilities. Having grown fat in manufacturing excellence during the years of plenty, German industry was shocked after 1990 to find itself becoming uncompetitive. Labour costs per unit of output by 1991 were 36 per cent higher not only than those of Japan, but of so-called *'verlorene'* Britain as well. They were 45 per cent higher than those in France.[11] New pressure has been felt from the low-wage work forces to the East, prompting one government official to comment 'It is as if we had Hong Kong just 80 km from Berlin'.[12] Meanwhile Germany's own former Communist *Länder* are costing DM150bn per annum, 100bn of it in benefit payments and wage subsidies. After 1995, nursing the former GDR will still cost the country DM120bn-130bn each year,[13] yet the country's magnet effect persists. While Kohl has proved capable of pandering to Germany's renascent nationalism, both in his reaction to the new xenophobia and his defence of German monetary pre-eminence, he will also be the longest-surviving architect of Maastricht and its frustrated agenda. The best that can therefore be said is that, by representing some continuity with the Bonn-Paris policy of Konrad Adenauer, he can give France more time to adapt to the changes overtaking this key axis.

For France's national interest now demands redefinition. The anticipated ejection of the Socialists from power at the March election constitutes the third milestone of 1993. Under the stewardship of Edouard Balladur, the new government has moved to implement policy decisions demanded in any case by the Community's agenda, such as privatising the state industrial sector and granting independence to the Banque de France. However, opposition to any reduction in agricultural subsidies—as much a test case for Europe's independence in Paris' view as a national economic concern—still unites all France's parliamentary parties, and the mood can only harden as the country slides further into the continent's recession. Furthermore, Balladur's function is that of caretaker until the President retires in 1995; by this time the 'grogne' in the economy is likely to make the next head of state—whether the Gaullist Jacques Chirac, the Europe-first Jacques Delors (Mitterrand's own preference) or even the accommodating Balladur himself—an uncompromising defender of national interests. German exasperation with France's obstruction to a transatlantic trade deal was being allowed to show through by late 1992 (Foreign Minister Kinkel stating 'We can tolerate many things, but not a trade war with the US')[14]; Bonn's preparedness even to override the Rome Treaty by giving preference to the bilateral agreement on telecommunications equipment[15] should be seen as a further warning.

Possibly, beset with the task of rebuilding a united country, Germans simply lack time to concern themselves with Europe-building, just as they were late in waking up to the gravity of the Gulf Crisis. The decay of the European vision has accentuated this psychological withdrawal; as the Danes prepared to give their second verdict on the Maastricht Treaty, some 66 per cent of German respondents hoped that they would reject it.[16] In a Community built around a Franco-German axis which imposed equilibrium on its two partners, this amounts to the lion taking control of its tamer. Nowhere is Germany's return to respectability more evident than in the defence of the Deutschmark. In Germany, the Mark is the safe, strong, never-devalued symbol of Germany's post-War development. Apart from being the cornerstone upon which the EC's other currencies rest, it also runs as a parallel currency across much of Eastern Europe. Kohl's attempts to buy his countrymen's agreement to sink it in a common currency—by making Germany the home of the European Monetary

Institute (EMI), its issuing authority—merely emphasise its supremacy.

The day when Western Europe enjoys a single united currency according to the Maastricht prescription has been put back beyond the horizon. This was the case even before the Treaty's criteria for economic convergence were sent sprawling to the point where only Luxembourg seemed likely to meet the agreed desiderata. While a number of countries moved to set their central banks free of government control (another EMU prerequisite), that this does not lead to bankers abandoning national policies is amply demonstrated by the *Bundesbank*'s interest rate policy. 'Increasingly improbable' is how Helmut Schlesinger, the *Bundesbank*'s president, described the 1997 target date for Currency Union,[17] as national performances were seen to diverge; his deputy remarked that an EMS with permissive loopholes would 'hardly be able to fulfil its purpose as an instrument of discipline'.[18] When, by August 1993, the currency grid had broken down precisely because of these gentlemen pursuing a national, anti-inflationary interest-rate policy, even Kohl allowed that the target date for Monetary Union might slip 'by a year or two'.[19] All the other EMS currencies bar the Dutch Guilder had by then been forced either out of their trading bands or out of the grid altogether, leaving the System all but dead in name. That Stage II of Monetary Union—the opening of the EMI—was envisaged for January 1994 is an indication of how far Europe has fallen out of step. That the *Bundestag* has given itself the last word on participation in the final march to monetary union in 1997 has become as irrelevant a technicality as whether Westminster gives the Sterling consent to participate. A former *Bundesbank* president, Karl Otto Poehl, expressed the widespread view of the whole Treaty baggage while waiting for the British parliament and Germany's Constitutional Court to deliver the final flourishes: 'My forecast would be that Maastricht is ratified, but that the probability of it ever being implemented is lower than ever'.[20]

Faced with the practical obstacles to unity in foreign policy, a uniformly vigorous defence posture, economic convergence and consequent monetary union, the most ambitious policy aims of Maastricht will still be in pieces at the end of the decade. In order to get the Treaty past Denmark's sceptical voters at its second attempt, it became necessary to prepare a special version, gutted of its remaining substance. Ostensibly delayed until 1994, the Schengen agreement on open borders has been sacrificed to French public opinion, the fear of immigration and the '*petit oui*' of 51 per cent France gave the Treaty in 1992. In fact polls suggested that public support across the Community for the unification process had fallen to an average of 73 per cent by May 1993 as against a high of 81 per cent at the time of the Treaty's signing.[21] The irony is that, having delivered the coldest shower, it is the Danes who have granted Maastricht its technical reprieve. (Denmark is one of only three Community countries where the Treaty even earns an absolute majority of approval.)[22]

Back to Basics

The Treaty, with the eventual ratification of Britain and Germany behind it, will thus be left in place as a signpost pointing towards the presently unattainable. Attempts to kickstart the process, as proposed by Europe's Christian Democrats[23] and the Belgian mid-1993 presidency of the Community, will continue to produce diminishing returns, addressing incidentals such as public transparency (as opposed to content), clarification of CFSP areas allowing majority decisionmaking, and the location of the irrelevant EMI. The next challenge is the absorption of Austria and the Nordics; while their governments may prove more federal-minded than Britain's Euro-pragmatists have hoped,[24] their electorates remain on their guard. At the turn of the century, the morphous Community will still be crawling hesitantly towards uncertain confederation. There will remain the Single Market, covering the whole of the European Economic Area, at best left as a pragmatic survivor of the reverse in the Community's fortunes in 1992-93. Pragmatic because implementation of the Single European Act of 1985 will remain subject to conditions until at least 1998, and because the lack of a uniform currency offers unending opportunities for indirect trade protectionism within the Community.

The wider prospect is of a weakening Alliance span without a united and resolute European buttress. Both the Anglo-American and the Franco-German partnerships have lost their dependability, and no European Union is yet poised to subsume the respective functions of the EC, NATO-Europe or WEU and European positions in the UN or G7. Meanwhile the existing collective and bilateral systems are both sapped by America's inward turn.

Where Europe will go from here, is anyone's guess. Clearly, despite all the difficulties, the EC is not about to unravel. Indeed, mainly in order to regain some credibility, all European institutions are now re-casting themselves, by considering the incorporation of East European states. Furthermore, despite violence on Europe's peripheries, the heart of the continent remains fairly calm, and is likely to do so in the future as well. Germany's current economic problems are unlikely to disappear soon. But they are also unlikely to destabilise German democracy, or the European order. Yet, at the same time, with a deepening recession and increasing political disarray, the temptation of governments to adopt seemingly easy solutions, such as restrictive trade measures in order to alleviate pressure on local beleaguered industries, is growing all the time.

The Prospect Of Trade Disputes

Trade disputes are almost indispensable ills which accompany any market economy. New products clamour for a fresh treatment; old declining industries use political pressure to survive; currency fluctuations and institutional changes can make matters worse in unpredictable ways. The world suffered a bout of trade rifts in the 1970s, another in the first half of the 1980s and is engaged in another one today. On every previous occasion, a trade war was avoided and economies expanded. Is there any reason to doubt a similar outcome today?

The answer is an emphatic *Yes*, for many reasons. First, because both the EC (with the possible exception of Britain) and Japan are sliding into a severe recession, just as America's recovery appears fairly shallow. Thus, negotiators from the world's main trading blocs have even less leeway than ever before to compromise. With wobbly governments everywhere, a premium is placed on any politician who offers a seemingly straightforward solution. Baiting the Americans (as France's now-defunct Socialists proposed) or kicking the Japanese in the teeth (as Ross Perot demanded during the American presidential campaign and ever after) is appealing to ever-larger numbers of electorates in democracies. The reappearance of the arguments that Japan is somehow special and must be treated outside the regular framework of trade relations, or that the EC's manifest destiny is to stand up to American 'bullying' (as Paris wanted to have us believe) are merely the most obvious manifestations of this trend.

Secondly, the end of the Cold War is now melting down all international cooperation institutions, which did so much to smooth over many of the former trade disputes. Should China be accepted into the GATT organisation? What is the role of the states of the former communist bloc in the world economy? What future for the NAFTA agreement and, indeed, for the European Community? Everywhere, all institutions are being tested, all are undergoing transformations and all are fighting for their survival. The turf battle between the IMF and the World Bank, or between the Americans and the Europeans over aid for Russia are some examples of this situation. The Uruguay Round of GATT is stalled not merely because of arguments about oil seeds, aircraft and cross-subsidies in other sectors. Instead, the very creation of new structures within this organisation is now intensely disputed. Institutions nursing their own wounds, defending themselves against alleged profligate spending (the World

Bank, EBRD) are not very good at defending the principles for which they were founded.

Third, the security situation around every important trade partner has now changed. The US feels that it has won the Cold War, but is in danger of losing the peace which follows. 'It's the Economy, Stupid' was the silly slogan on which a president entered the White House this year, a rallying cry for those who believe that the reason for America's ills is to be found among predatory foreign manufacturers. For Japan as well, the world now looks different. On the one hand, the military connection with the US is no longer viewed as essential in Washington, despite all the claims to the contrary from President Clinton. Yet on the other, Japan can no longer continue its traditional game of being a world economic Superpower, but a minion in security terms. The biggest arms race today takes place in South-East Asia, where the end of the Cold War is intensifying long-dormant animosities and reshuffling alliances. A nuclear North Korea? An aggressive China? A demilitarised Pacific due to withdrawing US troops? All such questions must be answered by Tokyo today. Finally, in Europe as well, all the games associated with the Community project are unravelling. Despite brave words about the Maastricht Treaty, European Monetary Union is unlikely to take place this century, and most Western governments know it. The Brussels bureaucracy still talks on behalf of 'Europe' despite the fact that it represents only 12 states on the richest, western tip of a continent waiting to be united. And the Single Market, launched with such fanfare in January this year, has studiously avoided such 'trifling' matters as liberalising telecommunications, frontier controls, insurance and banking services and air transport. All this has to be taken into account, even before one remembers China's phenomenal growth to world economic power status. Politicians bewildered by the pace of change at home are not very good international trade negotiators. And populations rattled by the sight of a melt-down in all the common assumptions with which they were educated for decades reciprocate accordingly. The glue of the Cold War, the automatic assumption that no Western government faced with the overwhelming security challenge of the Soviet empire would dare unleash a trade war, has gone.

With it, intellectual fashions have changed as well. It is suddenly academically respectable to talk of trade retaliation, of defending local economies, of supposedly easy cures for long-term ills. The academics who have been churning out all kinds of half-baked tracts about the need to protect America's economy are now in the White House. The Japanese, eager to accommodate claims on their economy even if Tokyo considered them unjustified, are suddenly determined to say *No* to any further demands. And the Europeans are caught somewhere in the middle.

In essence, never before since 1945 have so many balls been in the air, and so many of them still refuse to come down. Literally nobody is able to predict what will happen to GATT, Russia, Germany, Japan, the US, NATO, the UN. And trade cannot be separated from politics: one wrong step in one place could have a ripple effect on all other calculations. It is usually forgotten that trade wars do not start with a ringing declaration, or a motion passed in Congress. Almost every trade war in history was the result of either lack of international cooperation institutions (last century) or a wider political rift across the Atlantic (the 1930s). And every single one of them started with seemingly small, incremental steps. A shipping restriction here, a tiny retaliatory tariff sanction there: all ultimately contributed to a final disaster by launching a spiral of retaliation. The world trading system is today in incomparably better shape than ever before: finance has never been more international, and world institutions, although semi-paralysed, are still there. Nevertheless, the dangers of a wider trade conflagration are still looming and a bunch of bumbling politicians, all paying more attention to the latest opinion poll at home, can make it worse.

1. *The International Herald Tribune*, 1 February 1993.
2. US Information Service News Alert, 9 February 1993.
3. *The International Herald Tribune*, 13 May 1993; *The Independent*, 10 May 1993.
4. See, for example, Senator Joseph Biden, quoted in *The International Herald Tribune*, 12 May 1993.
5. See Jim Hoagland in *The International Herald Tribune*, 11 August 1993.
6. See for example *The International Herald Tribune*, 10 August 1993.
7. *The International Herald Tribune*, 4 June 1993.
8. Ambassador James F Dobbins to Centre for European Policy Studies, 24 May 1993.
9. Anthony Lewis, 'For America in Europe...', *The International Herald Tribune*, 25 May 1993.
10. *The International Herald Tribune*, 27 May 1993.
11. *The Economist*, 6 March 1993.
12. *The Financial Times*, 7 June 1993.
13. *The Economist*, 6 March 1993.
14. *The Independent*, 11 November 1993.
15. *The International Herald Tribune*, 14 July 1993.
16. Wickert Institut poll, published in *Frankfurter Allgemeine Zeitung*, 19 May 1993.
17. *The Financial Times*, 3 June 1993.
18. Hans Tietmeyer, quoted in *The Financial Times* 5/6 June 1993.
19. *The International Herald Tribune*, 10 August 1993.
20. *The Wall Street Journal*, 30 June 1993.
21. Eurobarometer poll, reprinted in *Libération*, 22 May 1993.
22. *Ibid.*
23. *The Financial Times*, 3 June 1993.
24. See *The Financial Times*, 11 June 1993.